LINKEDIN

LinkedIn

TELL YOUR STORY
LAND THE JOB

TYCHO
PRESS

CONTENTS

CHAPTER 5

Searching 63

How Keywords Work on LinkedIn 63

CHAPTER 6
Connecting 75

CHAPTER 7
Getting the Most Out of Your Connections 87

WHAT IS LINKEDIN?

What Is LinkedIn?

You don't head to LinkedIn to share the rowdiest pictures from your recent road trip, gawk at adorable baby animals, or gossip about the latest viral video. LinkedIn is an online refuge for career-oriented people interested in finding an amazing job, landing superb long-term clients, branding themselves, establishing a well-connected network of passionate industry authorities and experts, or a combination of these goals. In short, LinkedIn is an online social networking venue for businesses and career professionals (see fig. 1 for home screen).

The primary principle at work in LinkedIn is "the friend of a friend." It all begins with your "friends," or Direct Connections as they're referred to on LinkedIn. These Direct Connections are your "First-degree" Contacts. These "First-degree" Contacts, through LinkedIn's extraordinary service, open the way to your "Second-degree" Contacts—individuals that you could reach through the Direct Connections that you have in common.

LinkedIn Skeptics: Why Do You Need to Be on the Network Anyway?

LinkedIn is the ideal social network for the non-Facebook generation, or anyone who waxes fearful at the thought of releasing personal content online. The extremely professional and conservative formatting of the network gives LinkedIn the feel of a top-flight industry conference or stately business meeting, not the free-for-all other networks are evocative of.

Admittedly, many potential users are on the fence about joining yet another social network. Is there actually any value to establishing a presence on the site? Absolutely. Here are just a few compelling reasons to consider:

Sell yourself without selling at all. Being active on LinkedIn naturally exposes you to industry head honchos and decision makers. A substantial, well-worded profile is often all that is needed for valuable Contacts in need of your expertise to find you and reach out to you. LinkedIn makes it easy to enter the Discussions and Communities necessary to build and enrich your presence in the professional sphere.

Become a cutting-edge industry insider. What specifically would you like to learn more about in your current field or your dream industry? Can you name the top dozen experts in that field or industry? Which companies or products offer the most value in terms of your professional needs? What are the most important recent headlines relating to your career or business? Who's making this news? More importantly: who is your competition? Answers to all these questions live on LinkedIn.

Advertise your assets around the world, 24/7. A Google search for your name will often turn up your LinkedIn Profile—assuming you have one. Your presence on the network speaks to your reputation and potential value, even when you aren't around. LinkedIn is the perfect venue for

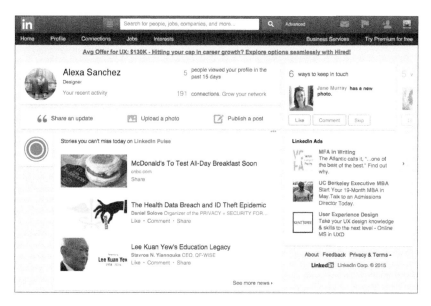

FIG. 1 Source: https://www.linkedin.com/

controlling exactly how anyone around the world can learn about you and make a professional connection with you.

Back up your reputation for career and business continuity. Nowhere else online can you more effectively trumpet your professional accomplishments and expertise than on LinkedIn. The network allows the Connections who know and value you best to sing your praises through enlightening testimonials.

LinkedIn Vs. Other Social Networks

Those new to the social media arena are often curious about how LinkedIn stacks up against its social network competition. Here's the lowdown on what makes this particular professional network stand out:

vs. Facebook: LinkedIn is often erroneously described by those unfamiliar with it as "Facebook for business." The comparison is not even remotely apt; the two networks are starkly different in their usability.

Whereas Facebook is great for casual encounters and relaxed web surfing, LinkedIn is meant for focused professional use—targeting and communicating with like-minded and career-oriented professionals.

vs. Twitter: The primary difference between these two networks is that very word, "network." Twitter does not require that you know someone for them to interact with you, or vice versa. But LinkedIn's "friend of a friend" structure requires a professional connection before communication begins. This creates an advantage for those who value developing a dynamic and supportive network, instead of a passive and one-sided one.

vs. Google+: The audience you're targeting will decide who wins this matchup. If you value ranking high on the world's top search engine, then a presence on Google+ will definitely be a priority for you. However, when it comes to building a targeted network of skilled professionals, either for a job searching or for generating sales leads, it's tough to top LinkedIn.

In the end, LinkedIn arguably trumps both Facebook and Twitter for many reasons. Your Profile is much safer on LinkedIn than on either of these other two networks, since you are allowed just one picture, and there's no "wall" on which others could tag you or leave potentially salacious comments. Many workplaces frequently ban Facebook and Twitter use, but LinkedIn's subdued professional aesthetic brands it as an essential work-related tool; even your boss is unlikely to bat an eye seeing your LinkedIn Profile on your screen. In a post-recession era, where job security is not guaranteed, it's all the more important to brand yourself and highlight your expertise. LinkedIn can help you define your professional worth, while at the same time organically attract potential opportunities for new jobs, networking, and sales—even while you sleep.

LinkedIn's Value for Job Seekers

Whether you're currently unemployed or not entirely satisfied with your current job, LinkedIn offers a wealth of tools and features to help you secure an amazing new job. Even if you currently find yourself in a great position, it never hurts to quietly explore new employment horizons. A great Profile on LinkedIn organically attracts recruiters eager to capitalize on your skill set, or even to poach you from your current job with a better offer. And you can advance miles ahead of coworkers through the high-level learning opportunities and network formation that occur in LinkedIn's Groups. No matter where you fall along the job satisfaction and security spectrum, a LinkedIn Profile serves as a long-term job security and career development Mecca.

LinkedIn's Value for Businesses, Marketers, and Sales Pros

LinkedIn provides an extremely effective format for attracting high-value clients and Contacts. If you're looking for a simpler way to generate consistent, quality sales leads, then you'll definitely want to invest time into developing an impressive presence on the site. Proficient LinkedIn use can significantly abbreviate your sales cycle, generate more sizable orders and purchases, and lengthen the lifespan of your client relationships. Staying "top of mind" becomes a cinch with LinkedIn, as does passively attracting deals.

Paid vs. Free Accounts on LinkedIn

Although the overwhelming majority of LinkedIn's best features and value are available at no cost, certain users may accrue additional benefits from upgrading to a paid Premium account (see fig. 2). Those just starting out on the site should probably avoid Premium for at least the first few months of LinkedIn use. Why? Because it can take that long to really get a feel for your individual, authentic voice, as well as establish a rhythm for how you'll interact on the site on a daily and weekly (and in some cases, hourly) basis.

Once you're comfortable enough with the site and decide to take the next step, what will you get with a Premium account?

LinkedIn InMails. When you are seriously interested in connecting with someone too far outside your network to message, properly crafted personal InMails can sometimes save the day.

More introduction messages. LinkedIn hands you 15 of these for free; aggressive job seekers and businesses who have amassed a network teeming with impressively-connected, willing Contacts (which can take months to develop) would benefit from this.

Deeper knowledge into who's checking you out. LinkedIn's "Who's Viewed My Profile" shows your most recent visitors, and only via a Premium account can you access every bit of this information. This is potentially of value to businesses looking out for both leads and rivals.

High-powered searching and visibility. Members with Premium accounts can search by coveted categories, like company size and

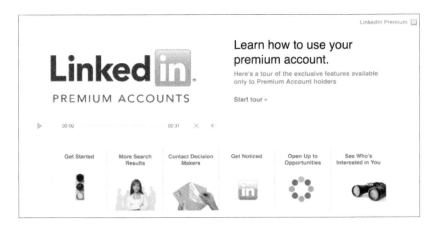

FIG. 2 Source: https://www.linkedin.com/

seniority level. In addition, Premium-account Profiles maximize what you can see in other members' Profiles.

LinkedIn offers you a free trial of Premium, which you should refrain from using until the network becomes second nature to you. After this, reevaluate how seriously you need these LinkedIn tools to fulfill your goals, and consider paying up for Premium.

The Top Nine LinkedIn Tools to Explore

Here's a sneak preview of the best features available to you on the network. While the chapters ahead will elaborate on each of these, this list should whet your appetite:

1. *Rich media.* A feature that many LinkedIn users don't take advantage of (but that you surely will) is the ability to furnish your Profile with credibility-lending videos, presentations, white papers, and other media representations of your professional talents. Adding even a very brief "hello" video to your Profile can dramatically increase the number of Connections that come to you—all for little additional effort.

2. *Headline.* The value of this feature is the immediate communication of what makes you valuable, both on and off LinkedIn. Including the right Keywords in your headline can organically and effortlessly draw attention from the exact professionals that you had been pursuing for years in the non-virtual world.

3. *Endorsements.* Endorsements are a quick and simple way to communicate your expertise in a particular area. Just be willing to delete those endorsements that aren't specific to your actual skill set, and prioritize those endorsements that capture all your career-based glory (see fig. 3).

4. *Groups.* Participation in those Groups most pertinent to your goals can significantly expand the reach and quality of your professional network. Your activity in Groups can serve as an extension of your

Skills

Top Skills

55	Distributed Systems	
49	REST	
48	Java	
42	Software Engineering	
38	Maven	
29	Open Source	
27	Software Development	
25	Agile Methodologies	
25	Git	
20	Web Services	

FIG. 3 Source: https://www.linkedin.com/

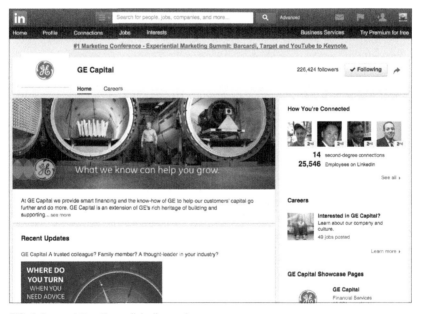

FIG. 4 Source: https://www.linkedin.com/

personal branding efforts, broadcasting your message to a highly desirable audience. They can also provide you with the intel and resources you need to be better at what you do.

5. *Headshot.* Don't worry—you don't need to be a model or movie star for this feature to benefit you. A professional headshot (with the emphasis on "professional") can instantly establish your character, credibility, and value, augmenting what LinkedIn can help you achieve.

6. *Recommendations.* They help you get into college, they help you land an internship, and on LinkedIn, Recommendations can open pathways to unique higher-level networking Contacts. The secret to making this work best: practice giving Recommendations as openly as you receive them.

7. *Company Pages.* Brand your business as an industry titan through this tool, available exclusively through LinkedIn (see fig. 4). It's the ideal way to become a thought leader and to passively generate leads to your service or product. In addition to a Company Page, LinkedIn also allows you to create impressive Showcase Pages for your wares.

8. *Advanced search.* LinkedIn is more than just a networking hub; it's also a high-powered, dynamic people- and prospect-scouting database. The network allows you to search for Contacts by industry, location, job title, or Keywords. If you have ever wished for a simpler way to find sales leads or the gatekeepers to your dream job, this feature is the answer to your prayers.

9. *Introduction requests.* Goodbye, cold-calling; so long, solicitation e-mails. The ability to get to know "friends of friends" through LinkedIn can make it so much easier to get in touch with a coveted professional who is just outside of your network, via your mutual Connections.

Now that you have a general sense of what LinkedIn can do for you, we'll drill down into some of the specifics in the coming chapters. In the next chapter, we'll take a look at your "face" on LinkedIn: your personal Profile.

A STRONG PERSONAL PROFILE

The wonders of LinkedIn in terms of finding a job or generating leads never cease to amaze. But without a strong Profile, tapping into such wonders will prove quite difficult.

Begin your journey on LinkedIn with sufficient time and attention devoted toward perfecting each detail of your Profile. Before starting, perform a quick search for other Profiles against which you might someday be matched: a recent college grad looking for a marketing job on the East Coast might search for "Boston marketing profile," for example. In order to establish a competitive Profile, think about what makes you, your brand, company, product, or service sensational. A great Profile is as substantial (i.e., heavy on content) as possible, features great specificity (through details and Keywords), and establishes credibility (via the numerous sections LinkedIn provides for this). This chapter will walk you through the steps of easily creating such a standout LinkedIn Profile.

A Tour of Your Personal Profile Page

Lets take a look at what a typical LinkedIn Profile looks like (see fig. 5). You will first see the Essential Information box on your Profile page, which displays your name, profile picture, location, industry, and headline (at the most basic level, your education, the company you work for, and the position you hold there). In the bottom-right corner of this box, you can view the number of Connections you currently have. On the bottom-left corner, you can see the full public URL that links to your Profile.

The Background Section below this box provides more specific information about your professional record. It covers your Experience (which, essentially, is your resumé), your Summary (a vital section that builds on your Experience), your Education, your Skills & Endorsements, and your Honors & Awards. This section can be further customized to include the languages you speak, your volunteer experience, and other dynamic details.

Recommendations, a very important but often overlooked section, follow your Background information. You'll next see the Groups to which you belong, and the Following box, which details the influencers, news sources, and companies that you are tracking on LinkedIn.

On the right-hand side of your Profile, you'll notice the Profile Strength meter (see fig. 6), which evaluates how substantial your Profile currently stands. You'll rank as a Beginner just by creating a Profile, but following this book's advice will easily push you into the Expert or All-Star levels.

You can also check out Who's Viewed Your Profile. This section shows you how many visitors your Profile has received in the last 90 days, as well as your relative percentile ranking for overall Profile views compared to your LinkedIn Connections. Certain users choose to keep their Profiles private, meaning you won't be able to see their information. Choosing to purchase LinkedIn Premium will reveal past viewers to a much larger extent, but it won't override a visiting Profile's anonymity.

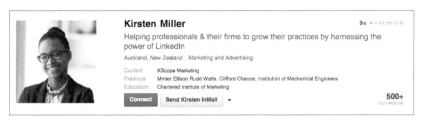

FIG. 5

Profile Strength

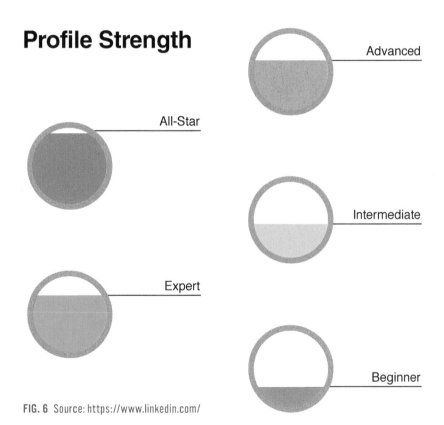

Advanced

All-Star

Intermediate

Expert

Beginner

FIG. 6 Source: https://www.linkedin.com/

The Start to an Exceptional LinkedIn Profile

You are now familiar with the essential structural components of a LinkedIn Profile. But how you specifically go about compiling each component will communicate volumes to those who may visit your Profile—recruiters, valuable networking Contacts, potential customers, and the like. Here are some indispensable insights into maximizing the effectiveness of your Personal Profile, with a focus on the value of each component.

What's in a Name?

The most straightforward Profile component can sometimes pose a few challenges. Be sure to include the most relevant professional or academic certifications and acronyms for your career here, such as MD, PhD, Six Sigma, or CPA. You can incorporate your maiden name parenthetically, like Raven (Lee) Johannsen. If you have a nickname, former name, or more extensive maiden name, include it as a Former Name (move mouse to the top of your homepage over Profile, and click Edit Profile. Click your name, then hit Former Name). Finally, select your visibility restrictions and Save.

Make Sure the Camera Loves You

LinkedIn fosters a spirit of professionalism and communication, making it imperative that you have a picture on your Profile. Not just any photo will do; you'll need a professional headshot (see fig. 7). Such a picture shows you in flattering professional attire, avoids a distracting background (i.e., a loud color or annoying pattern), and captures that "I'm an excellent candidate ready for hiring and networking" look in your eyes.

LinkedIn has recently offered users the option to "Add a background photo," an option you are strongly advised to take advantage of. No other Profile feature allows you to brand yourself and maximize your web

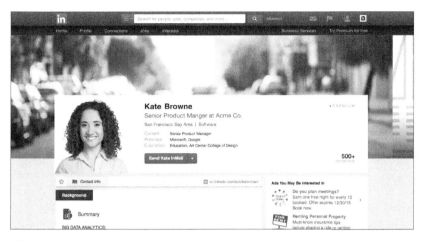

FIG. 7

presence's visual appeal as much as the 1400 x 425 pixel background photo. Collaborating with a graphic designer on this image is the ideal option, but free or cheap image-creation services like PicMonkey and Fotor can also accomplish the task. Color choice is crucial, as is content. For example, if you're a noted guest speaker, incorporate shots of you at the podium into the background image; to give another example, a childrenswear retailer should favor primary colors over muted pastels.

Writing a Headline That Pops

Your 120-character headline (see fig. 8) provides a smart opportunity to let your profile visitors know who you are, and entice them to start a conversation with you. Here's how to construct it correctly:

What value do you offer?

Consider the problems you solve, or the reasons that you deserve to be hired. Job seekers should not only state that they are "seeking employment," but also the specific position they are most interested in securing. Businesspeople and entrepreneurs need to communicate, in the headline, the essence of why customers need and should want their product or service.

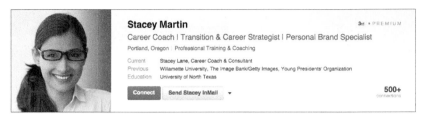

Stacey Martin   3rd • PREMIUM

Career Coach I Transition & Career Strategist I Personal Brand Specialist

Portland, Oregon : Professional Training & Coaching

Current   Stacey Lane, Career Coach & Consultant
Previous   Willamette University, The Image Bank/Getty Images, Young Presidents' Organization
Education University of North Texas

Connect   Send Stacey InMail ▾

500+
connections

FIG. 8

Keyword selection.

Avoid generic words like "executive" or "specialist," and concentrate instead on the specific phrases that distinguish you. An owner of a floral shop, for example, should consider inserting the words "wedding flowers" or "bridal bouquet pro" into their headline.

Save space and add character(s).

Use the "pipe" key (Shift + Backslash on the keyboard) to free up additional room. A quick Google search for "LinkedIn symbols" will expose you to the various smileys, check marks, and other images that you can copy-and-paste into the headline as well.

Location, Location, and . . . Industry!

LinkedIn will automatically determine your location based on the zip code you enter when creating your Profile. Those who live just outside of major cities should always opt for the larger municipal option instead of the exact town that matches your zip code; doing so will broaden your search range.

Take time to consider the most appropriate industry for your Profile. Your best industry will reference your company's sector rather than your specific job category: If you are working in the HR department of a major bank, it would be wiser to select Banking as your industry rather than Human Resources.

How Contacts Get in Touch with You

Click the Contact Info tab just below the Essential Information box in your Profile to view exactly how visitors will be able to reach out to you. Take a look at your e-mail address; if there's a more appropriate address than the default option LinkedIn automatically selects, enter and save it now. You can also link your other social media profiles and websites here. Note that only your First-degree Connections will be able to view all of this information. Everyone else will only see your website(s) and Twitter info.

LinkedIn does a great deal to essentially serve as your own tailored professional website, and the fact that you can customize your public Profile URL demonstrates this well. (To customize this, mouse over the hyperlink to the left of the Contact info button, click on the pen, and you'll be directed to your public Profile. Glance at the right-hand side and scroll down to "Create your custom URL"; ideally, this should be something that will resonate with your actual name, making it simpler for visitors to remember and access your Profile.)

The Value of Your Projects on LinkedIn

Include relevant Projects on your Profile in order to demonstrate your authority and expertise in your industry and particular position. Job seekers will find this a handy location to prove their aptitude to those who may look to hire them, while businesses can include value-based projects that educate their target customer and encourage visitors to want to learn more about a particular product. If you collaborated on a Project with team members, by including them here each of their names will automatically become a hyperlink that directs to their LinkedIn Profiles.

Privacy Settings: Can You Be Too Public?

Remember that your public LinkedIn Profile can be accessed by anyone on the web. You can opt to minimize the amount of information that can be viewed by both public visitors and users of the LinkedIn network. LinkedIn's popularity practically ensures that your public Profile will surface on Google first-page search results, which can be troublesome to those who aren't interested in web-based attention. On the other hand, this lends an advantage to savvy professionals and job seekers. (Access your privacy settings by mousing over your Profile picture and clicking "Privacy & Settings." You can also "Customize Your Public Profile," controlling what elements show up on search engine results, by clicking on your public Profile URL link and looking at the right-hand side of the screen.)

Creating your Profile with as professional an intent as possible—meaning, omitting any nonprofessional information about yourself—allows for anyone on the web to gauge valuable information about you, ranging from adequacy for a potential job or networking opportunity, to expertise that could net a possible client. Packing your personal Profile with substantial yet professional content maximizes LinkedIn's effectiveness for furthering your aims both inside and outside of the network.

Beefing Up Your Resumé Through "Experience"

It's best to view your entire LinkedIn Profile like a dynamic, super-charged CV; the battery that powers this CV is the Experience section (see fig. 9), which also is the Profile component most similar to a traditional resumé. When considering what jobs to include or leave out of this section, evaluate each potential inclusion through the following lenses:

Credibility: Every job you mention should either underscore your particular expertise in your industry, give a potential recruiter or networking Contact confidence in your aptitude for a new position, or convince a customer to trust in your experience and savvy regarding your product or service.

Experience

Publications Manager
The American Strudent Dental Association
January 2014 – Present (7 months) | Chicago, IL

I manage the print publications as well as in-house design projects for the association. As staff liaison to the editorial board, I work with 8 dental students to plan content and find authors for ASDA News (newsletter 10x/yr), Mouth (quarterly journal) and Mouthing Off (blog, 3x/week). I also work on the following:

- Maintain brand standards across the association
- Design, layout and final production for periodicals and all print collateral.
- Manage the RFP process fro print projects and printer relations.
- Manage the publications section of the association's website.

FIG. 9 Source: https://www.linkedin.com/

Noteworthiness: Is the job you're thinking of adding to your Profile a position or skill you are well-known for? Remember that LinkedIn functions as perhaps the world's most impressive professional search engine. If a job's title or description contains an asset you'd want to be searched and singled out for, include it.

Competition: Take time to scout through LinkedIn and survey the landscape of your professional peers. Would someone looking to hire an expert in your field prefer your Experience to someone else's, or vice versa? If a particular job gives you an edge over other Profiles, add it.

The Importance of the Summary Section

If the Experience section is LinkedIn's equivalent of a resumé, think of the Summary section (see fig. 10) as your Profile's cover letter. Arguably the most important component of your Profile, the Summary provides a 2,000-character opportunity to brand yourself, promote engagement from Profile visitors, and significantly improve your odds of landing a job, sale, or client, or of making a connection. After your Profile Picture and Essential Information, the Summary section is the very next component that your LinkedIn visitors will see, underscoring its importance.

Summary

Even though I'm a PR person by trade, I'll always be a reporter at heart. I'm incapable of pitching something I myself don't wholeheartedly believe in. I have a passion for uncovering unique and compelling ways that people are using a product, service or site and revel in knowing that I can help thousands of other people hear those stories.

Currently I'm LinkedIn's Senior Manager, Corporate Communications (Mobile Product PR Lead). During my years at LinkedIn I've had the privilege of working in several communications roles including product, corporate, consumer, trade (advertising and HR) and even international (Brazil, Canada and Japan) public relations. I travel quite a bit, so I have a knack for living out of one bag of an extended period of time (event when I fly to multiple climates :).

Stephanie Rosenbloom did a lovely job writing up a few of my favorite pieces of travel advice in this article in The New York Times, "How the Tough Get Going: Silicon Valley Travel Tips" http//lnkd.in/nyttravel

Specialties: Adding a human element to corporate stories, helping editors/reporters get great angles, international PR, managing PR firms so they feel like a partner not a vendor, setting goals, and surpassing them, social media, being a solid spokesperson, broadcast news, television, TV, radio, newspaper, public relations, media relations, reporter, reporting, press, speaker, panelist, keeping my cool under pressure and being someone you can rely on when you're on a tight deadline.

FIG. 10 Source: https://www.linkedin.com/

Here are some tips to help your Summary become the space to tell your story so that it highlights your value in the best possible light.

Think "story," not "speech." The best LinkedIn Summaries enchant Profile visitors with their narrative structure and possess a strong beginning, middle, and conclusion.

- ***Start your Summary (300–1,000 characters) by elaborating on the specific value you provide: job seekers should concentrate on clarifying the elements of the positions they're best suited for, along with***
  the top one or two specific reasons they are a great fit for these positions. Businesses should highlight the specific product or service they provide, who would most benefit from it, and how exactly they would benefit.

- ***The middle section (500–1,000 characters) should build on the Summary's beginning.*** Elaborate on your qualifications. What have you done for previous customers, employers, or positions? What jobs in the Experience section would you like to comment on a bit more that makes you stand out? If you're moving into a new field, how does your previous experience qualify you for a new industry?

- *Conclude your Summary (300–1,000 characters) with ways Profile visitors can learn more about your achievements or projects, or what valuable relationships you have with other LinkedIn members.*

Break it up. One chunk of blocky text is not as visually appealing as two to five shorter, more readable paragraphs. Use headers, sub-headers, or graphics to divvy up your Summary and fuel interest in the section.

Use every single character. Skimping on your Summary is ill-advised. Don't forget about the "pipe" key (Shift + Backslash) to help you create more space. Low on content? Add additional Keywords. Provide more Contact information. Include a quote from a testimonial that you couldn't add to Recommendations. Creatively work your way up to the 2,000-character maximum.

Additional Profile Sections to Explore

Besides Experience and Summary, LinkedIn has an array of other sections to lend your Profile valuable substance. Here's just a sampling of additional worthwhile information you can add to your Profile:

- Any *Languages* you have mastered
- *Volunteer Experiences* that could attract an employer
- *Volunteer Opportunities* you're searching for
- *Organizations* to which you belong, like the American Marketing Association or the Public Relations Society of America
- Noteworthy *Certifications* (CPR, Better Business Bureau, etc.)
- *Causes* and *Supported Organizations*, in categories like poverty alleviation or disaster relief, that cast you in a positive light
- Impressive *Test Scores* you've earned
- *Patents* that you have secured for your innovations
- Specific academic or professional *Courses* that distinguish you

- Particular *Interests* you have that might attract employers or customers

One final tip: be aware that you can easily reorder nearly every element of your Profile. (Click the icon named "Drag to rearrange profile sections" that appears in the title box of every section.)

Don't Forget to Include a Recommendation (or Two, or Three)!

Say your Summary sells you well; your Recommendations, coupled with a strong Summary, are what will incentivize an employer, customer, or client to buy into your Profile. And just like other Profile elements, every Recommendation is Keyword-searchable. No LinkedIn Profile is complete without at least three to five solid Recommendations that vouch for you, at least one for each job you mention in your Experience. Job seekers will want to target current or former teachers and professors, employers, and mentors, while business professionals should aim for networking colleagues and highly satisfied clients to sing their praises. Here are some tips on securing great Recommendations for your Profile:

Prioritize Contacts with great LinkedIn Profiles. It's best to secure a Recommendation from someone whose own LinkedIn presence you respect and wish to emulate.

Chat with the individual recommending you beforehand. By phone or over e-mail, express your gratitude for their willingness to support you, and communicate your request that they discuss your value with adequate detail and substance.

Look the Recommendation over before posting it to your Profile. You may need to edit or revise a Recommendation to best promote your value and credibility. Thoughtfully and politely make these suggestions known; these changes should happen without trouble.

Skillfully Select Your Skills & Endorsements

LinkedIn is notorious for e-mails and notifications regularly sent to users to endorse their Connections for seemingly outlandish or irrelevant skills. Although Endorsements can seem rather purposeless and generic, you can reconfigure how you utilize Skills & Endorsements in order to maximize its potential to support your skill set.

Do not list or include every single skill you have for potential endorsements. Consider what would be the 5–10 most relevant skills you have that meet two criteria: your Profile's ability to verify that skill (through Experience, Summary, or an Additional Section), and the value of the skill to your target Profile visitors—potential employers, recruiters, networking Contacts, clients, or customers in your particular industry or occupational sphere.

Any skill that doesn't satisfy both of the above criteria should immediately be removed from your Profile. Those skills that remain will not only ring true to your Profile, but should also organically attract abundant endorsements from your LinkedIn Connections. (You can edit the order of your Skills or Endorsements by clicking on one of them in the section and looking down toward "Drag to reorder.")

Maximize Your Profile with Media

The finishing touches to a truly accomplished LinkedIn Profile lie in the media you select (see fig. 11) to turn up the volume on your presence in the network. An overview of the most valuable media you should add follows.

Publications. Do you maintain a blog that shows your expertise through its every post? Are you a regular guest blogger on important sites? Have you written a white paper, book, or article that communicates what

 Publications   + Add   ↕

The Unofficial Book On HootSuite: The #1 Tool for Social Media Management
✎ Edit   ↕
The Social Media Hat / Amazon / Barnes & Noble

May 25, 2014

The Unofficial Book On HootSuite provides business owners and social media managers an in-depth look at the most popular social media management tool. It offers clear instruction, recommendations, and power tips. It is available on Amazon for Kindle and on Barnes & Noble for Nook.

How to Manage Your Google+ Brand Using HootSuite   ✎ Edit   ↕
Plus Your Business!

February 11, 2014

This article details how brands with a Google+ Page can use the social media management tool HootSuite to post and monitor and engage.

Warning: This will improve your blog traffic by 25%   ✎ Edit   ↕
Business2Community

January 30, 2013

Article outlining the methods and benefits of using stronger blog titles, and the impact those titles can have on website traffic and social media engagement.

First Step in Newsjacking: Monitor News and Industry Trends   ✎ Edit   ↕
Social Media Today

January 26, 2013

Article discussing Newsjacking as a blogging technique, and the use of Google Currents as a news monitoring tool.

Google unveils powerful marketing tool: Think Insights   ✎ Edit   ↕
BizMarketing Magazine

May 6, 2013

Article on Google's new Think Insights marketing data platform has been published in the online magazine, BizMarketing, in their May 2013 issue (registration required).

FIG. 11 Source: https://www.linkedin.com/

makes you a true professional? Do not hesitate to add your published work to this section for an extra boost.

The "Add Media" icon. Scour your files for PDFs, PowerPoint or Slideshare.com presentations, and YouTube videos—anything that can help Profile visitors visualize your credentials. If you don't have any relevant media at hand, try to create at least one piece. Even a 30-second introductory video or quick screenshot of your latest blog entry can illuminate your accomplishments. (To add media, head to the Summary or Education sections, or to any job in your Experience, and look for the icon that reads "Click to add a video, image, document, presentation . . ." on the right-hand side of the section title box.)

Final Thoughts

Let's summarize the various components to establishing your best Profile on LinkedIn:

- Work on a great Profile picture and headline.
- Prioritize your Experience and Summary sections.
- Secure multiple Recommendations.
- Beef up every valuable Additional Section you can.
- Personalize your Profile with a custom URL, a correct professional name, working Contact information, and access to your relevant links other than LinkedIn.
- Incorporate several pieces of media into your Profile.

In the next chapter, we'll look at ways to get the most out of Company Profiles.

COMPANY PAGES

Company Pages allow you to share valuable information about your organization, products, services, and even in-house career openings. In addition to a company website, a corporate presence on LinkedIn provides a rich opportunity to expand and nourish a brand's presence.

Is a Company Page Right for Me?

If you're looking to maximize high-quality exposure for your business, consider it essential to have a LinkedIn Company Page. Company Pages can drive targeted attention to the value that you offer to potential clients and customers. Company Pages can also help you build a subscriber base, as LinkedIn provides the option for pages to earn "followers," a self-selected audience to whom you can send updates, news, and invitations to connect.

Searching for a way to spotlight specific products and services? A Company Page can support this goal as well, through "Showcase Pages" that allow you to promote your business in a natural way.

Finally, the Recommendations feature on Company Pages allows satisfied clients to give you valuable testimonials, all on their own accord. Simply put, Company Pages serve as a powerful tool to magnify the impact your business creates online.

Company Page vs. Personal Profile

Although Company Pages carry the potential of worthwhile soft-selling and relationship-building with possible clients, they are not necessarily the most effective choice for every business. Clients and Contacts are more likely to willingly interact with an individual than a corporation, and significant work is required to lend a Company Page the "soft sell" atmosphere that is essential to making it successful. Although Company Pages can endow your enterprise with followers, you can't use them to proactively create conversations with other LinkedIn members in the same way as you can through a Personal Profile, since Company Pages can't participate in Groups or send individuals direct messages.

On the other hand, Company Pages do provide a forum in which all employees can connect at a single dynamic, virtual hub. Company Pages can also pay for Sponsored Updates, allowing you to promote certain pieces of content to a targeted audience. And as with a Personal Profile, Company Pages can be outfitted with targeted Keywords and phrases, improving their search engine value.

In sum, a Personal Profile doesn't require a Company Page in order to thrive on LinkedIn. But using a Company Page without combining it with a well-established Personal Profile can result in diminishing returns in the network's value for your business.

Company Page Basics

In order to create a Company Page, be sure to address all of the following:

- Ensure that you already created a Personal Profile, complete with your actual first and last name.
- Fully complete your Personal Profile so that it reaches an "Intermediate" or "All-Star" level of Profile strength.

- Verify that you have approximately five or more First-degree Connections on your Personal Profile.
- Check that the company for which you would like to launch a page is listed in the Experience section of your Profile; you must be working at this company in a current position.
- Add and confirm your company-related e-mail address to LinkedIn, ensuring that the domain name of the company is unique and individual. (Gmail, Yahoo!, and similar e-mail services cannot be used to start a Company Page.)

How to Create a Company Page

Once you've met all the criteria necessary for starting a Company Page, you can create one by following these steps:

1. Mouse over *Interests* at the top of your homepage and select Companies.

2. Click *Create* in the *Create a Company Page* box on the right.

3. Enter your company's official name and your work e-mail address.

4. Click *Continue* and enter your company information.

   - *If the work e-mail address you provide is unconfirmed on your LinkedIn account, a message will be sent to that address.* Follow the instructions in the message to confirm your e-mail address, and then use the instructions above to add the Company Page.
   - *A red error message may appear if you experience difficulties in adding the Page.*
   - *A preview of your Company Page is not available.* Once you publish the page, it goes live on LinkedIn.
   - *Finally, you must include a company description (250–2000 characters including spaces) and company website URL.*

How to Add Content to a Company Page

Launching a Company Page is only truly complete after you have added specific information on the value your business can provide through your services and offerings. As long as you are the Administrator of the Page, you have permission to add this information. Follow these steps to create spaces on your Page for a maximum of 25 business offerings:

- Click the Products & Services tab on the Company Page.
- Head to "Edit" at the top right, and click "Add product or service," then "Get started."
- Select the product/service you'd like, along with its category. Name the product/service as well.
- Include a description, key features, a website, a named Contact, an image, and (if possible), a descriptive video.
- Hit "Publish" at the top right.

Adding a Company Page Banner

To make sure your Company Page really pops, add a branded banner. First, make sure your banner image meets the following dimension requirements: minimum 646 x 220 pixels, PNG/JPEG/GIF format, maximum 2 MB, and a landscape layout (image should be wider than it is tall). When in doubt, go for a slightly larger-size image, as smaller files will not upload to LinkedIn Company Pages. Next, head to your Company Page and click "Edit" at the top. Look for "Image" or "Logo," and hit either "Edit" or "Add image." "Upload" the image and "Save," then "Publish."

Careers Tab

The Careers tab is an excellent feature for businesses that don't mind paying for the service. It allows a Company Page to interact with potential job seekers, both actively and indirectly searching for employment opportunities. Many companies employ this tab to great effect (see examples at the end of this chapter).

Showcase Pages

For particularly unique or standout products, services, or initiatives your business offers, you should consider highlighting them on a special Showcase Page; a feature that will allow you to flaunt this offering to an interested audience. LinkedIn members can Follow both your Company Page and any Showcase Pages that may capture their attention as well (see examples of well-done Showcase Pages at the end of this chapter).

How to Create a Standout Company Page

An excellent Company Page manages to achieve two feats simultaneously: building brand awareness and trust, while also maintaining consistent and meaningful interaction with Page visitors and followers. Here are some strategies you can implement to best manage this balance:

Field a Company Page Go-To Team in Your Company

Having a group of two to three talented employees who can lead the upkeep of the Page provides an enormous advantage to the long-term usefulness of LinkedIn's network for your business. Try developing a round-robin editorial calendar, where each team member handles a particular set of tasks (updates, content creation, etc.).

Employ Excellent Branding

Populate your Company Page with your business-based images, banner, and logos. Graphics like these lend all-important brand recognition to your Page, helping customers remember your service or product over the long-term. Make sure that every graphic you have is highly legible, is professionally produced, and catches the eye of the viewer.

Inspire Engagement with Your Followers

Ask questions and respond to all comments. Consider hosting an on-brand contest or sweepstakes on your Page.

Define Your Ideal Customers and Their Needs

Before beginning to round out your Company Page in earnest, return to the drawing board regarding your buyer persona; refresh your memory about the specific needs, pain points, and characteristics that distinguish your target client. Be sure that every element of your Company Page is, in small yet potentially meaningful ways, honoring and catering to your target audience in a detailed and thoughtful manner.

Take Advantage of Video

The importance of video for brand-building has swelled in recent years. Wherever you can, take the time to include video-based content onto the Company Page. Create a 30-second welcome video that greets visitors when they head to the Page. Import company-created short videos from Vine or Instagram. Encourage past satisfied clients to create video testimonials for your products and services. Host a weekly or monthly interview series with different members of your business team. Leverage your network to develop Q&A videos, themed on the problems your services address and solve, with industry leaders you know.

Create a "Follow Us" Button for Your Company Website

Attract more attention to your LinkedIn Company Page by adding this feature, with the help of LinkedIn's Developer Page (developers. linkedin.com—search "build a follow button").

Examples of Great LinkedIn Company Pages, Showcase Pages, and Careers Tabs

While you continue to flesh out your Company Page, take a glance at what some businesses are doing. Sure, some of these organizations have sizable marketing budgets to draw from. Still, even the smallest business can take lessons from these examples to improve its own Page.

Company Pages

Pacific Dental Services. This company enjoys ranking as one of the top dental companies in the country, thanks in great part to its exceptional Page. The admin staff here meticulously addresses all comments and creates connections between followers.

Mashable. Regularly featuring on Company Page best-of lists, Mashable stands out for fine-tuning its famously widespread content base to the precise needs of its LinkedIn audience: integrating social media and pop culture themes with advice on business development and productivity (see fig. 12).

NPR. This media corporation stands out for its strongly-shaped content offerings. Rarely will you notice long, overdrawn posts here. Instead, NPR LinkedIn followers enjoy short, appetizing lists and "How To" pieces, creating a stream of consistently high-value content.

Four Seasons. The famed hotel group accomplishes two brand-building goals at once on their Page: presenting Four Seasons as a must-apply-to organization for hotel management and luxury professionals, while also encouraging LinkedIn businesspeople to take a much-needed vacation.

FIG. 12 Source: https://www.linkedin.com/company/mashable

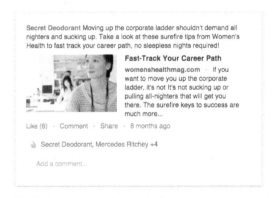

FIG. 13 Source: https://www.linkedin.com/company/mashable

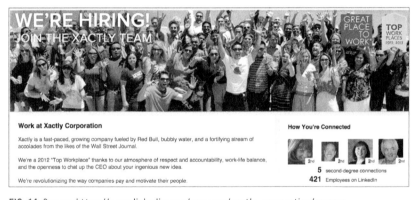

FIG. 14 Source: https://www.linkedin.com/company/xactly-corporation/careers

Showcase Pages

Secret Deodorant. Procter & Gamble entices and educates its target customer with a smorgasbord of content on the popular deodorant brand. Notice how every post elevates the overall discussion beyond just anti-perspirants, and into conversations on office attire and work-life balance (see fig. 13).

Adobe Creative Cloud. You would do well to emulate this page's use of powerful and evocative imagery in branding. The expert graphics on this Showcase Page create a cohesive experience for Page visitors, underscoring the connection between the product and its purpose, all the while maintaining a very professional front.

Careers Tabs

Luxxotica. Every element of this tab encourages you to apply immediately to work for this company, from the well-produced welcome video to the behind-the-scenes style banner and informative slideshow.

Xactly Corp. Their concise writing style maintains viewer interest and stimulates greater engagement with tab visitors. It provides a great example of valuing potential employees' time while underscoring what would make Xactly such a great fit (see fig. 14).

With your Personal Profile shining and your Company Page in place, it's time we turn toward some of the more social aspects of LinkedIn. In the next chapter, we'll look at ways you can take advantage of Groups for professional success.

GROUPS

LinkedIn Groups 101

LinkedIn provides Groups as valuable opportunities to get in touch with other individuals in your location, industry, or specialization. Many users cite Groups at the most worthwhile component of the entire network. Once you've got a fully-furnished Profile, consider Groups the next best extension and use of your interaction time on LinkedIn.

Why Do Groups Matter on LinkedIn?

Groups provide multiple sources of overall professional value to users by:

- Allowing you to connect with like-minded LinkedIn community members
- Promoting important events and activities that could benefit your career
- Providing a way to find you through the network
- Offering an opportunity to listen in on the needs of potential customers and steal nifty tips from fellow industry pros

- Posting employment listings and Discussion topics for job seekers
- Providing a way to talk to Group members who are outside of your direct LinkedIn network

Suffice it to say that ignoring Groups is ignoring the full power of LinkedIn.

What Can I Accomplish Through LinkedIn Groups?

The sky is the limit when it comes to what Groups can do for you. Depending on your particular reasons for using LinkedIn, the network's Groups can accommodate and fulfill your goals. Below is just a sampling of what's possible for your LinkedIn experience with some smart Group participation:

- Become a thought leader in your particular field
- Brand yourself or your business and substantiate your credibility
- Locate and engage potential clients or customers
- Keep up-to-date with the latest and most important trends and happenings
- Network with top-flight industry professionals who might otherwise be inaccessible to you
- Build an impressive base of knowledge and insight in a given sector or conversation
- Find your dream job
- Maintain a high-activity Personal Profile (through frequent group interactions), a quality that quietly attracts potential employers and recruiters

How Many Groups Should I Join?

The jury is out on the exact number of Groups you would be recommended to join. Many experts claim that joining as many as you can is best, but it's hard to believe that maintaining an active presence in each of those Groups (an important goal) would be possible in the long-term. Consider joining a number of groups most relevant to your particular needs that you would be able to participate in on a regular basis, or at least two to three times a week. (Examples of active participation would include providing a valuable post, answering a question, or interacting with another group member.)

Should I Receive or Opt Out of Group E-mails?

The more Groups you join, the more e-mails you'll be receiving; this number can grow rapidly depending on the size and activity level of each Group. Larger Groups are more likely to send you mostly irrelevant messages, meaning that turning off e-mail notifications for such groups might be a stress-relieving move. But be sure to keep receiving messages from smaller groups, as well as groups whose subject matter you relish. Job seekers may want to maximize their notifications for Groups that regularly discuss employment postings, and businesspeople scouting potential clients may find monitoring Contact-rich Groups to be a smart strategy.

Which Types of Groups are Right for Me?

Not all groups are for everyone. Use the following criteria to decide whether or not a group will suit your needs (see the end of this chapter for a list of popular groups):

By location. Search for Groups that cater to your particular region, especially if you are located in a more remote part of the country. For example, Ohio-based users would be well advised to favor groups targeting Cincinnati or Toledo, while it wouldn't be practical for a Brooklyn-based user to join every Group in New York City.

By specialty. This is a no-brainer, as long as you narrow down your target. Joining every marketing Group would not be as smart as participating in groups specific to your markets.

By location and specialty. Combine the first two criteria for a range of Groups that could satisfy you. An HR professional in San Diego, for example, should search for HR Groups at the local (San Diego), regional (Southern California, California, and West Coast), and national levels.

Organic affinity groups. Never fail to join alumni associations, hobby groups, groups based on spiritual affiliation, and other Groups that resonate with your identity and distinctiveness.

LinkedIn User Groups. Since LinkedIn regularly updates the network with improvements and structural adjustments, joining these Groups will help you keep up with the latest strategies for effective LinkedIn usage.

How Do I Find My Best Groups?

The first step in finding Groups best suited to you involves combing through the LinkedIn Groups Directory, where many (but not all) of the network's most interesting Groups are listed. Then, consider the following questions to help you locate the most suitable Groups to join:

Where are the best jobs or clients for you?

Search LinkedIn for Keywords related to your ideal position or potential customers, as those will be prime locations for even more of what you're after.

Where are your competitors interacting?

Conduct a search for fellow companies, users with Profiles similar to yours, or users who already have the job or clients you wish you had. See which Groups they're affiliated with, and join them.

LinkedIn Group Use Basics

Finding a great selection of Groups is the first step toward professionally engaging with new communities. Read ahead for strategies that will maximize the advantage each Group can provide you with for the time you spend there.

The First Three Steps After Joining A Group

After spotting and joining your ideal LinkedIn Groups, you may want to immediately jump in and begin interacting with fellow members. While you're absolutely free to do so, following these three steps will improve your Group contributions and ensure you get the most value from your participation:

1. *Listen.* Take time to read through the Group's rules, Discussions, and comments from members. This will give you a good idea of the right style in which to reach out.

2. *Help someone.* Have you recently read an article that could clarify a Group question? Do you have someone in your network that could be of service to a Group member? Make your first post about lending a helping hand in this way, and your efforts will be rewarded.

3. *Comment thoughtfully.* Build on the first two tips with conscientious contributions. Tailor what you write with specific and pinpointed phrases and ideas. Your attention to detail will make you a Group standout.

Posting to LinkedIn Groups

Posting to a Group is as simple as posting to your Personal Profile; just look for the box with the phrase "Start a Discussion or share something with the group . . ." This box is essentially your Discussion post title, and can only span up to 200 characters. Once you start typing, the option to "Add more details" will appear; elaborate on your Discussion post title here. You can also specify if your contribution is of a general nature or job-related. Hit Share, and you're off to the races. All Group members, except those who have turned off notifications, will receive your post in a group e-mail. In each post, try to strike up a conversation with other members. Use an inviting or stimulating question to title your post, and always respond promptly to any comment you receive; these tactics will help to keep your post high in the overall Group Discussion feed.

Conduct Successful Group Searches and Find the Most Valuable Group Members

You can directly interact with members in a Group that might not otherwise be in your LinkedIn network. Each Group, therefore, is a potential bounty of Contacts that could prove useful in your bid to win a great job or land a long-term client. Understanding how LinkedIn Group searches work is the first key to mining Groups for Contacts.

Experienced Internet users have probably heard of and used Boolean search terms. Fortunately, LinkedIn Groups utilize the same terms. Essentially, these terms allow you to specialize and narrow down the exact type of results you seek from each search. There are three types of search options:

"NOT." Use this to remove a certain term from your overall search. For instance, if you wanted to find recruiters in any field except marketing, you would type in "recruiter" NOT "marketing."

"AND." This type of search comes in handy for two-term requests. Say you want to find Group users located in Houston interested in insurance. You would enter "Houston" AND "insurance" into the search bar.

"OR." Employ OR searches when results of either search term will satisfy you. In pursuit of Group Members located in either New York City or Philadelphia, you would enter "NYC" OR "Philadelphia."

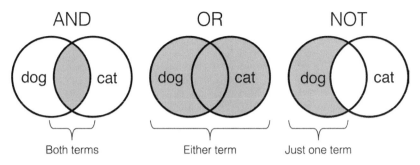

FIG. 15

To take LinkedIn Group search to the next level, click Search, then enter the location(s), keyword(s), and/or job title(s) that you would like to select from that Group. The members that result from this search are your new target Contacts.

Either scout their Profile, or message them directly through the Group (a great feature!) about your interest. You can employ this strategy to mine Groups for sales leads, valuable networking Contacts, or employers and recruiters you'd like to draw to your Profile.

Tools for Finding Great Group Material

It's not enough to simply make any post to a Group. Ensure that every post you make contributes value—a quality that could improve your online reputation and drive traffic to your Profile. Here are three great resources to help you find awesome content for your Group posts:

Pocket. This app is a perfect web-surfing companion (see fig. 16). Whenever you come across an intriguing article or video that could be catnip for your Group, use Pocket to save the content in a convenient, organized format.

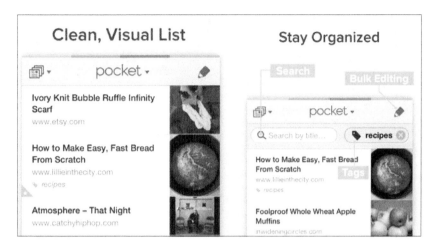

FIG. 16 Source: https://getpocket.com

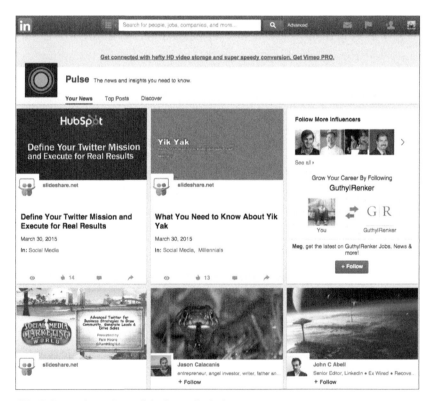

FIG. 17 Source: https://www.linkedin.com/today/

Feedspot. Never again fall behind on your blogs and favorite websites, since Feedspot gathers the latest posts from all of your favorite online sources in one neatly arranged feed (and, if you like, a daily digest e-mail.)

LinkedIn Pulse. One of the latest features on the network, this resource aggregates all the best content from your LinkedIn community members in one handy page (see fig. 17).

Should I Start My Own LinkedIn Group?

Launching a Group of your own could prove a nifty way to attract tailored job leads and a steady stream of potential clients and new colleagues. But if you think participating in Groups is time-consuming, just imagine the workload required to start and maintain your own Group for the long term. It takes serious, sustained effort to make your Group worthwhile. Spend at least a few months regularly using LinkedIn as an active participant before considering creating your own Group. Make sure that your Group's topic is both unique (not covered by established Groups) and attractive (potentially thousands of LinkedIn members might be interested). Before taking the plunge, brainstorm the value your Group would offer, and come up with fifty potential Discussion post ideas, to thoroughly evaluate your Group concept.

The exact protocol for starting a LinkedIn Group changes periodically, but the basic steps are pretty much self-explanatory. In the Groups page, look for the option to Create a Group. From there, fill in the required info and, voilà, you have a Group. Beyond that, develop the template messages Group members will receive. Here are the types of messages you can craft for your templates:

- **Request-to-join Message**—for people who request to join the Group.
- **Welcome Message**—to people when you approve them for membership in the Group.
- **Decline Message**—when you decline requests to join the Group.
- **Decline-and-Block Message**—when you decline requests to join the Group and block any future requests.

Using LinkedIn Groups for Great Results

Now that you've been schooled in what makes Groups great, you're ready to benefit from even more tailored information on strategic Group participation. Both job seekers and businesses can exceed their professional expectations with LinkedIn Groups. Here are some tips on how to leverage Groups to discover a career-changing industry connection, win an amazing position, or snag a dream client or sale.

How to Use Groups for Amazing Networking

Again, the Groups feature is what distinguish LinkedIn as the prime spot for professional networking online. Take advantage of this with the following relationship-oriented tips, any one of which could one day have you communicating with a top-level Contact in your field.

Ask and answer one question every day.

If you do both of these on a daily basis, you will quickly earn a reputation as someone who deserves to be Contacted for rewarding conversations outside of the Group. Asking a pertinent, stimulating question communicates your passion and curiosity about the Group's topic, while answering an existing question colors you as both helpful and an expert.

Connect daily with one Group member.

Regularly following the above tip will eventually encourage someone to reply or thank you for your posting. This creates the perfect opportunity for you to message this person, set up a phone conversation, learn more about their professional background, or—best of all—earn a new First-degree connection, which expands your LinkedIn network exponentially.

Refrain from promoting yourself.

While this may sound counterintuitive, always err against touting your own horn; gear everything you write to the specific subject at hand, or the person you're networking with. Too much first-person commentary ("I," "me," "my") will mark you as a self-promoter, or, in other words, someone more interested in selling himself than in connecting with the Group.

How to Use Groups to Find A Job

As with every function on LinkedIn, a fully fleshed-out Profile is a must in order to distinguish yourself for employment. Once you've assured yourself that your Profile represents your qualifications in an excellent manner, follow these tips to maximize Groups' effectiveness in your job hunt:

Always click the "Jobs" tab, a convenient feature of every Group.

The Jobs tab is not just geared toward employment-related Discussions. Do not mistake Groups' Jobs tab for the larger one that appears at the top of all pages on LinkedIn. Those looking for great, targeted new hires will often post new job information only in Groups. This narrows down their search to particular Groups, in order to tap the richest talent pool.

Utilize advanced Group searches.

What are the top companies or organizations for which you'd like to work? Conduct searches to find Discussions specifically related to these companies. Prepping for an interview with your dream workplace? Search through your groups to find relevant info in advance of the interview, a move that can generate great ideas for shaping what you'll say and the questions you'll ask the interviewer.

Become an ideal candidate through research.

Groups can yield excellent information on internships, volunteering experiences, industry-specific associations and events, certifications, and other potential assets that could distinguish your resumé. Mine Groups for opportunities that will make you stand out to employers and recruiters.

Add valuable Discussion posts and comments on a regular basis.

Be sure to contribute one to two thoughtful Discussions a week, and three to five detailed and pertinent comments on other Group posts each day. Doing this will maintain your Profile's visibility within the Groups you participate in, making it easy for recruiters to spot your conscientious commentary—and potentially message you for an interview.

How to Use Groups to Generate Sales Leads

Because of the professional trustworthiness that characterizes LinkedIn, the network has developed a reputation for being a community open to and passionate about new business opportunities—as long as those opportunities are relationship-oriented. Users who sell themselves don't succeed on LinkedIn, but users who provide value and listen to potential clients relish the lead generation LinkedIn offers. Here are a few tips for capitalizing on possible leads in Groups:

Prioritize Groups where your potential customers are.

What professional needs does your target client most complain and fret about? What are the specific interests and values that make your ideal buyer distinctive? Utilize this information to search for and join your most client-friendly Groups. Doing so will position you for both listening to customer pain points and, when the time is right, extending a helping hand with your product or service.

Offer a helpful, non-sales-focused, educational resource.

Develop a content- and value-rich e-book, webinar, podcast, infographic, or white paper that your target customer will love. Post this content as a Discussion topic in relevant Groups, making sure to use a pertinent question as a post title, one that a potential client would ask herself about the topic of your content.

Never, ever, sell in a Group.

LinkedIn members roundly snub any and all sales- or promotion-related content or writing styles. Getting caught pushing your product or service could lead to your being removed from the Group altogether.

Groups to Consider Joining

Whether hunting for a job, potential clients, or networking Contacts, LinkedIn's Groups are a true treasure trove. Below are some of the absolute best ones to consider joining. Please note: Private Groups require you to be approved by the Group leader, meaning that your Profile will probably be evaluated for relevance to that Group before you gain entry. Please adhere to all Group rules, and never spam or spoil the LinkedIn Group experience with your contributions.

The 14 Best LinkedIn Groups for Job Seekers

A Job Needed, A Job Posted. The Group's straightforward title emphasizes its practical, single-minded focus—to connect those searching for employment with those looking to fill vacancies, with as little muss and fuss as possible. (Private)

Executive Suite. This Group provides its members with an information-packed regular newsletter, in addition to providing you with access to recruiters, employment coaches, corporate executives, and industry-leading business experts. (Private)

Job-Hunt Help. Any question you could potentially have about conducting a job search both on and away from LinkedIn can be answered here with aplomb. (Private)

Career Explorer. Recent college graduates, as well as students a year or two out from graduation, will particularly benefit from this Group's opportunities, especially its college-specific subgroups. (Private)

Encore.org. Those who already have serious work experience under their belts, and who are looking to transition into a second long-term career, should beeline to this Group. (Open)

Talent HQ. Keep pace with the latest shifts and strategies in interacting with potential employers and recruiters through this Group's informative posts. (Open)

Career Rocketeer. Few LinkedIn Groups offer such a high level of active Discussion and member-to-member support as this one, which aims to help you simplify your job-hunt both online and offline. (Open)

Jobs DirectUSA. If you're looking to concentrate on landing a job as quickly as possible, consider prioritizing membership to this Group, whose users give advanced tips on employment-searching success through social media. (Open)

Linked:HR. This Group concentrates on Human Resources jobs and networking opportunities. Mine the posts here for rare tips on floating to the top of the HR talent pool. (Open)

MyCredentials: Career Presentation. Maximize your resumé, both on your Profile and your physical copy, with the help of this Group's selection of career and networking prep authorities. (Open)

Job Openings, Job Leads, and Job Connections! The largest LinkedIn Group, this resource gets to the point immediately; it focuses on juicy job listings from a myriad industries, as well as on forging connections between Group members on both sides of the job-hunt. (Open)

Career Change Central. Ready to move into a completely different professional field? Begin your transition by joining this Group, where you'll find career coaches, networking Contacts, and—yes—recruiters and employers in abundance. (Open)

The Undercover Recruiter. Bask in awesome exposure to hungry recruiters and employers worldwide through this Group; the direct-from-recruiter tips you'll glean on Profile and job search strategy improvement make membership here worthwhile. (Open)

Self-Recruiter. Instead of exposing you to recruiters, this Group endeavors to convert you into a targeted employment-seeking machine, dramatically improving your odds of landing an excellent next position. (Private)

The 15 Best Groups for Entrepreneurs and Business Professionals

Band of Entrepreneurs. Legal advice, PR savvy, and wisdom for every step of the entrepreneurial process, directly from a Group of caring and passionate innovators. (Open)

Small Biz Nation. Owners of small-to-medium enterprises should consider this a fount of valuable knowledge and networking opportunities. (Open)

Entrepreneurs Meet Investors. Members of both sides of the funding conversation make deals and make merry in this Group. (Open)

B2B Marketing. Seriously improve your aptitude in luring businesses interested in your particular product or service through this Group's exchange of wisdom. (Open)

Start-Up Phase Forum. Entrepreneurs and small business owners literally just beginning to make their mark can move ahead rather quickly by implementing what they learn in this Group, and by interacting with potentially long-term networking Contacts. It's a valuable Group for work-at-home professionals and home businesses as well. (Private)

Innovative Marketing, PR, Sales, Word-of-Mouth & Buzz Innovators. Hundreds of insight-packed daily Discussions justify this Group's lengthy title. The only thing bigger than its name is the pool of extraordinary marketing and PR authorities with whom you could be networking right away thanks to this group. (Open)

E-Commerce Network. Running an online business? Discuss every relevant topic under the e-commerce sun, alongside impressive top names in the industry, in this Group. (Private)

Content Strategy. Every business needs to create great content: white papers, blog posts, presentations, e-books, webinars, and the like. If you aim to improve your content-creation expertise, or to brand yourself as a content-strategy consultant, this is your top Group to join. (Open)

On Startups —The Community for Entrepreneurs. The single largest Group for entrepreneurs on LinkedIn's network maintains an intimate feel, with its vocal and passionate members discussing financing, hiring processes, and much more. (Open)

Young Entrepreneur Collections. Teen and twenty-something business professionals and innovators should earmark this Group as their top pick, with its penchant for helping young pros interact with consultants and top-level business execs. (Open)

Small Business Marketing Network. Build your skills in spreading the word about your small business, locate a marketing specialist, or even consider promoting your own marketing expertise in this active Group. (Open)

Women's Network of Entrepreneurs. Female business professionals and professionals-in-the-making find support and tailored expertise and networking opportunities here. (Private)

Let's Get Funded. If you are determined to get the seed money required to launch your startup or new business venture, aim to regularly participate in this Group's wise and practical Discussions. Angel investors, venture capitalists, intermediaries, and more, all flock here to lend their expertise. (Private)

NonProfit Marketing. Those LinkedIn users focused on building up a nonprofit enterprise should regularly immerse themselves in the expertise that distinguishes this Group, hitting every component of the nonprofit marketing process. (Private)

Business Development. A no-brainer join for every tech-focused business professional reading this book, this Group will build your skills in lead generation and sales. (Private)

Now that you have a handle on the purpose and function of LinkedIn Groups, it's time we looked at how to best navigate the immense amounts of information that exist on the network. Toward that end, the next chapter focuses on getting the most out of your search efforts.

SEARCHING

Creating a strong Profile or Company Page is the best preparation to step up your use of LinkedIn. The key to harnessing the network's incredible search functions lies in optimizing the use of very important descriptors—known as Keywords. Great Keywords can lead you toward an awesome job, track down an amazing client, or forge a lucrative networking Connection.

How Keywords Work on LinkedIn

Let's revisit the essential purpose of maintaining a presence on LinkedIn: to utilize your professional credentials toward the ends of networking, finding employment, finding clients, and/or building a personal brand. All of these purposes rely on the development of a presence—ideally one that is easily found on LinkedIn. On this network, there is an unmistakable correlation between the quality of your Keywords—the terms and phrases that capture who you are and what you want (see fig. 18)—and the quality of the results you obtain from your LinkedIn experience. Developing and implementing a set of strong, targeted Keywords will dramatically improve how well LinkedIn works for you.

Paul Burger I Therapist 3rd
Web Design

Therapist Website Design I Coach Website Design I
Chiropractic Website Design I Massage Therapy Web
Design

Greater Denver Area · Marketing and Advertising

Current   Brighter Vision: Therapist Website Design, 717 Marketing Inc,
           Glacier National Park Travel Guide

Previous   The Mandelbrot Project, Examiner.com, Breakthroughs

Education   Throw in some juicy keywords

Connect   Send Perry InMail   ▼   **444** connections

☆   Contact Info   www.linkedin.com/in/boulderseoconsultant/en

FIG. 18

How to Keyword-Optimize Your Profile In Two Steps

1. *Utilize audience perspective to determine your Keywords.* If you're
   pursuing a job, think about the types of phrases and terms recruiters and employers would use to find an ideal candidate for a position
   in your industry. If you're pursuing networking Contacts or clients,
   consider what Keywords these types of people rely on to find a product, service, or organization like your own. You can research this by
   making lists of the top 10 Profiles of similar job seekers or business
   leaders, or your ideal customer, networking Contacts, and thought
   leaders; jot down the dozen or so phrases you feel best identify each
   Profile; then survey the phrases or words that crossover between
   Profiles. Those phrases will serve as the base of your own Keyword
   list. Also, consider your industry, location, and job or professional
   title as additions to this list.

2. **Work Keywords into the most important Profile elements.** The Headline, Experience, and Summary sections really require as many Keywords as possible, provided that you incorporate them in a way that feels natural, not in overly technical "keywordese." Also consider adding Keywords into your Contact Information. If possible, edit your Recommendations to include your Keywords where appropriate.

How to Keyword-Optimize Your Company Page in Three Steps

1. **Look to your rivals and your dream clients for Keyword ideas.** Similar to the first step in Keyword-optimizing a Profile, the process of effective Keyword use for a Company Page starts with some serious research. Take time to investigate your competitors' Pages; what Keywords do you notice most? Definitely steal the most appropriate of them for your own list, but also think about developing your own Keywords that position you as a better alternative, or Keywords that are aimed at a slightly different target buyer. On that note, try performing Advanced Searches on LinkedIn to determine the dominant terms and phrases that appear in the Profiles of those LinkedIn users you could envision as ideal customers.

2. **Add Keywords throughout your Company Page overview, Showcase Pages, and Careers Tab.** These three Page elements characterize your entire business on LinkedIn, making them natural priorities in the Keyword optimization process. Remember, however, to incorporate your Keywords in ways that would make sense to a customer visiting your site for the first time. If it sounds too technical or forced, don't hesitate to remove Keywords to make for better sense and readability.

3. *Post a Keyword-charged status update once or twice a day.* This strategy is multifaceted in its benefits to your Company Page. The repetition of these Keywords in daily status updates will reinforce the phrases as a component of your business's brand awareness, helping you stand for these Keywords in the minds of potential new customers. This also forces you to consistently post on LinkedIn, which expands your reach on the network and also serves as an impressive traffic generator to your Company Page and website.

Keyword Optimization for Job Search

Remember to think of your Profile on LinkedIn as a dynamic, fully searchable, web-based resumé. In this respect, this book has so far recommended adding as many sections as possible to your Profile as are relevant to your professional characteristics and assets. You can take this a step further by Keyword-optimizing additional sections for the best search-friendliness on the network. Here is a grouping of the most productive areas to add strong Keywords on your Profile:

Languages. You can be descriptive here: is it "fluent" or "conversational" Spanish? "Basic" Russian, or "native" Cantonese?

Location. If you live in Honolulu and want a job in the city, "Honolulu" and "Hawaii" would be key search terms. Living in Honolulu yet job-searching for Minneapolis, however, would prioritize "Minneapolis" as the better Keyword.

Job titles. The job you're in now, and the jobs you've worked previously, are all prime areas to optimize with Keywords.

Name-dropping. Did you have a well-known former employer? Was your adviser in college a Nobel Prize winner? Do you consider a best-selling author in your industry a strong networking Contact? The names of such important people should be included as Keywords.

Job-specific terminology. What are the unique tools and techniques of the trade that you can bring to the industry in which you'd like to work? What certification, software (including desktop and mobile apps), or hardware typifies the field? Include these as Keywords as well.

Company Search

Searching for a specific company on LinkedIn can serve as an excellent ice-breaker for both job-seekers aiming for an interview, businesses looking for competitors or potential customers, or anyone looking to build high-level networking Contacts. This is simple to achieve. Simply enter the company you're interested in, and search for a First- or Second-degree Connection you share with someone working at the organization. Simply send a friendly message, or reach out for a warm Introduction.

How to Search for and Nurture Potential Customers

Simply put, the better your relationships on LinkedIn, the more success you'll experience in consistently attracting targeted and long-term clients to your product or service. Here's a selection of tips to help you improve your Connections and land better customers:

Warm up your First-degree Connections. A bad habit many LinkedIn users struggle with is adding "a bounty of new Connections without taking measures to build and maintain a strong relationship with each one. Remember that LinkedIn works by degrees of separation; the network allows you to connect with "friends of your friends" through Introduction Requests. However, your First-degree Contacts won't be so willing to let you tap into their network if they feel like you're using them. Prevent this scenario by keeping in touch with your First-degrees on a regular basis.

- *Message them with questions **that frame them as experts, as the providers of solutions to your most important queries.***
- ***Get offline.*** Schedule quick catch-up chats by phone or in person over coffee or lunch. Better yet, find networking events where you could both participate, and invite them to join you.
- ***Spread the relationship to other networks.*** LinkedIn is distinctly professional, of course, so adding your First-degrees on Facebook, Twitter, Pinterest, or Instagram can allow you to keep in touch in a more casual setting.

Look to the Newsfeed. Don't make the mistake of overlooking LinkedIn's front page in favor of your Groups, Personal Profile, and Company Page. The Newsfeed can frequently feature excellent possibilities of Contacts to add to your clientele. If you scroll down on the homepage, you should eventually see a feature entitled "See anyone you know? Connect with them": this is a hotbed of leads to explore. If you spot that one of your First-degrees has recently connected with one of the Contacts high-lighted in this feature, reach out for an Introduction Request. This is an intelligent way to pinpoint "warm" Connections on LinkedIn, which will make the likelihood of converting that Connection into a sale that much higher.

Identify the decision makers. This is an extraordinary way to mine your Contacts for potential leads in companies that you feel are a strong match for your product or service. First, develop a list of the top 25–50 companies you would like to connect to your business. Next, search your First-degree Connections for anyone who might be an employee at one of the companies on your list. After this, send a message warmly informing them of how your product or service could help their company; ask if they might know the company Contact in charge of decisions on purchasing. If provided with this information, you now have a platform to continue a value-based correspondence directly with the Contact—one that could eventually convert into a sale (See the final tip in this section for more on turning sustained communication over time into sales).

Send a carefully-crafted invitation message or InMail. It all begins with a well-executed Advanced People Search, the details of which can be found in the "How to Search for a Job" section of this chapter. (Instead of searching by job titles, search using a list of the top 10-20 Keywords that identify your target clients' pain points, identifying characteristics, and values that would direct them to your product or service.) If your Search goes particularly well, do not hesitate to Save the search, so that you have it on reference as a potential top lead-generating tool for your business for the long term.

After having identified a potential customer or client, you'll next need to determine whether it would be better to connect with them via an invitation message or through LinkedIn's paid network e-mail service, called InMail. If you have a "warm" connection to this particular lead (through a Group, a shared First-degree Connection, the same alma mater, or a mutual networking Contact), choose the invitation message option. If your lead is "cold" (meaning there is no shared Connection between you), but you still strongly feel that your product or service would be of value to this Contact, and you are willing to incur the expense of a subscription to one of LinkedIn Premium's paid account features, an InMail would be best.

In the case of an invitation request, lead with the details of how you both connect. In the case of an InMail, be sure to include a value-based point of potential connection: send them a presentation, e-book, or webinar from your company that would specifically address their needs, for example. In both cases, always lead with value, and refrain from any selling or promoting to any degree. The point of this well-crafted message is not to land the one-time sale, but to potentially establish long-term or repeat-business correspondence with this client; which leads us to the next tip . . .

Build relationships over time with value, thoughtfulness, and consistency. Establishing trust and confidence over a period of time best forges strong relationships with customers. Modern-day customers do not want to be sold to; rather, they prefer to feel that in making a purchase or committing to a deal, their intelligence and savvy made the sale a wise decision for them. In this respect, it's your job to foster this fidelity with clients. As previously mentioned, regularly providing clients with value-based media (e-books, webinars, infographics, slideshows, informative blog posts and articles, etc.) can win their trust. You should also strongly consider establishing a pattern of regularly following up with customers, which you can easily do through the LinkedIn Contacts feature (Contacts.linkedin.com). Remember to never sell or promote in your communication with clients; always center your messages on value for the customer, not your brand.

How to Search for a Job

Searching for a job on LinkedIn has everything to do with leveraging your First-degree Connections for potential Second-degree employment goldmines. Make sure that your First-degree Connections are individuals that you know in person and that you trust. Ask yourself, "Would I be willing to Contact this person in order to access his or her own Contacts? And, if so, would doing so pay off in forging a new connection?" If the answer to both of these questions is yes, then this First-degree Contact is the ideal person to add to your network.

Next, perform a search on LinkedIn for "recruiter in [enter your industry here]." Immediately message First-degree Connections about your availability for employment. Regarding the Second-degree Contacts that appear in the search results, you have a great option for connecting with them through LinkedIn. Go to the Profile of the Second-degree Connection, click the Down arrow, and select "Get introduced," which will pull up any First-degree Connections you share in common. Write a warm and friendly message to the First-degree Contact on why getting in touch with the Second-degree professional is important to you.

And always take a look at the "Jobs You May Be Interested In" and "Jobs In Your Network" modules that appear both in the Jobs section of the site as well as along the right side of your Personal Profile. This is LinkedIn's way of attempting to organically connect you with jobs that meet your particular skill set, industry, or serious interests.

Openly Hunting for a Job?

Those who are ready for great employment today—and are currently unemployed or unabashedly passionate about searching for a new job—should take full advantage of the space for optimization that LinkedIn's Profiles provide. Here are some ideas to ensure your openness for a new position is trumpeted loud and clear:

Do not hold back on your headline.

Be as specific, detailed, and colorful as possible in this Keyword-rich Profile component. "Office Assistant Looking for Work" is far too generic, uninspired, and passive to be effective on LinkedIn. What would be vastly better: "Highly Detail-Oriented Office Assistant Determined to Bring Productivity and Peace to a Needy Workplace." Yes, it's long, but this headline is much more likely to catch—and keep—the attention of recruiters and employers (see fig. 19 for some fiery headlines).

Run an ad.

If you are truly determined to win a job through LinkedIn, the network can provide you with advertising space targeted specifically at employers and recruiters you feel might be most interested in you. Be advised—it may take a few hundred dollars for your ad to run for the two to four weeks it can take before receiving a steady stream of responses. And before running an ad, make sure that your Personal Profile has been fully optimized (with Recommendations, a substantial Summary, and media examples of your skill set in action) to ensure that those who click on your ad are inspired to get in touch.

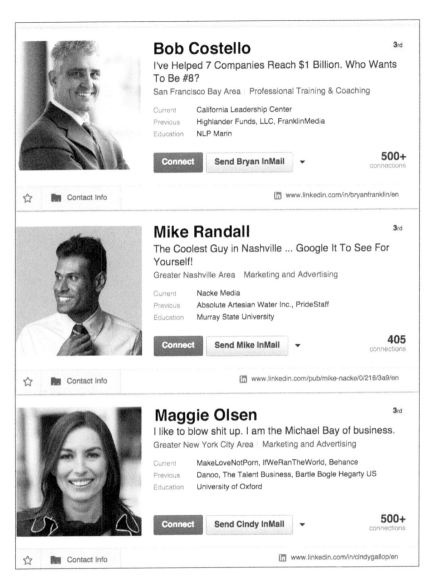

Bob Costello

3rd

I've Helped 7 Companies Reach $1 Billion. Who Wants To Be #8?

San Francisco Bay Area | Professional Training & Coaching

Current   California Leadership Center
Previous   Highlander Funds, LLC, FranklinMedia
Education NLP Marin

Connect **Send Bryan InMail** ▼

500+
connections

☆ ▦ Contact Info   in www.linkedin.com/in/bryanfranklin/en

Mike Randall

3rd

The Coolest Guy in Nashville ... Google It To See For Yourself!

Greater Nashville Area   Marketing and Advertising

Current   Nacke Media
Previous   Absolute Artesian Water Inc., PrideStaff
Education Murray State University

Connect **Send Mike InMail** ▼

405
connections

☆ ▦ Contact Info   in www.linkedin.com/pub/mike-nacke/0/218/3a9/en

Maggie Olsen

3rd

I like to blow shit up. I am the Michael Bay of business.

Greater New York City Area | Marketing and Advertising

Current   MakeLoveNotPorn, IfWeRanTheWorld, Behance
Previous   Danoo, The Talent Business, Bartle Bogle Hegarty US
Education University of Oxford

Connect **Send Cindy InMail** ▼

500+
connections

☆ ▦ Contact Info   in www.linkedin.com/in/cindygallop/en

FIG. 19

Job Searching Under the Radar?

There is no such thing as a truly "permanent" position anywhere. Taking the initiative to scout for potential better opportunities elsewhere will help ensure your professional longevity for the long haul. Understandably, you probably don't want your current manager to catch wind of your looking for greener pastures. LinkedIn, thankfully, provides a "quiet-search" strategy that brings new meaning to "one in the hand is worth two in the bush."

Be sure to track your dream companies through the Follow feature. Develop a list of your top 10–20 ideal companies and organizations, search for their Company Pages on LinkedIn, and Follow. Doing so will instantly fill your inbox with any employment notification or availability that comes up for these organizations.

Advanced People Search

After having taken all of the previous measures, use LinkedIn's exceptional search engine to ensure you've come across every possible appropriate job lead. Here's a work-through of exactly how to use Advanced People Search (see fig. 20) for brighter employment horizons:

1. *Click the "Advanced" button by the search magnifying glass icon near the search bar.*

2. *One by one, enter the top five to ten Keywords that capture the essence of the specific job title you are looking for.* Alternately, you can search for the title of your academic degree, which will yield a number of relevant ideas for potential jobs that match your educational background.

3. *As a final option, enter the names of your fellow alumni, co-workers, or networking Contacts at a level in their career similar to or slightly higher than your own.*

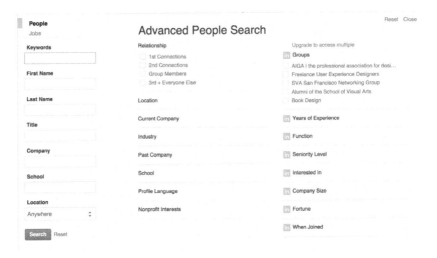

FIG. 20 Source: https://www.linkedin.com/vsearch/p?adv=true&trk=advsrch

4. *Hit "Search" to see a list of targeted professionals interested in the skills, educational background, or Connections you indicated.*

5. *Investigate these Profiles to learn about the details of their current and recent employment.* Also, snoop around to find out which Groups they are members of.

6. *Using either a direct message, an Introduction Request, a message through a Group, or an InMail.* Get in touch with these professionals and indicate your interest in employment opportunities that they may know of.

The LinkedIn Job Search App

Nearly half of all job seekers on LinkedIn are making use of this extraordinarily useful application for their employment pursuits. Here are the top reasons why you should consider joining them:

Undercover job searching. All of the activity you conduct on LinkedIn via this app is private and will not be shared with anyone, or be accessible on your LinkedIn website Profile.

Inspiration to keep up the hunt. Searching for a job can seriously drain your energy and test your stamina. This app makes it simple to continue your search with ease, no matter where you go.

Real-time notifications. The app allows you to receive the good news about, for instance, recruiters visiting your Profile. Even while away from your computer, you're intimately connected to your potential employment success via the app.

Feel like you have a handle on search? Good. Now, it has been said that you're only ever a handshake away from that next big opportunity. Thanks to the wonders of social networking, LinkedIn makes virtual handshakes a lot more common, and a lot easier to get. In the next chapter, we'll look at best practices for connecting and building up your LinkedIn network.

CONNECTING

While the value of LinkedIn may have its roots in the quality of your Profile, it eventually bears fruit as a result of Connections made. Read along to discover how to make the absolute best of your Connections for your job search, lead generation, or networking efforts.

How to Import My Contacts to LinkedIn

LinkedIn makes it easy to add Contacts from other online resources, like the address book attached to your e-mail account. On your LinkedIn Profile, head to the Contacts tab. From there, look for "Add Connections." Add your e-mail address and password, and then "Continue."

To connect with someone you know, but whose e-mail address you don't have, simply type their name into the search bar in pursuit of their LinkedIn Profile. Once at the proper Profile, hit "Connect," add a personalized message, and send.

When Connections you make from either method join your network, they officially become First-degree Connections with you.

Whom Should I Connect With On LinkedIn?

Always consider new potential Connections on LinkedIn the way you'd see someone calling or e-mailing to get to know you, or meeting someone in person. Would you respond to that call or e-mail? Would you want to continue the connection with that person after meeting? Then consider that Connection a valuable one to add to your network.

Here are some questions to always ask yourself before officially allowing someone into your LinkedIn network:

- Do I value this Connection's First-degree Connections (who would become your potential Second-degree Contacts)?
- Would I feel comfortable asking this person for introductions to Contacts in their network?
- Am I intrigued by the professional opportunities made possible via this Connection (potential resources for a job search, improving my networking base, a potential client or connector to potential clients)?

Of course, you probably shouldn't accept every request to connect that you're offered. Here are a few reasons to reject a LinkedIn Connection Request:

- When you and the potential Connection have never worked together before, or collaborated with one another in a meaningful, professional way
- If you feel as if the Connection is unable to vouch for you professionally, or vice versa
- If the requester sent you the default "I'd like to add you on LinkedIn" message, refusing to take the time to personalize her message to you
- If you suspect that this Connection request might be an attempt to annoy you with spam, often indicated by the lack of a Profile Picture

A great option to consider, if you're on the fence about fully accepting a Connection, is the Reply feature. Your invitation will show an Accept button, with a downward arrow next to it. Click that arrow to select "Reply (don't accept yet)." Ask how the requester knows you. Replying in this manner is a great option for Connections you make through Groups and via other social media outlets that eventually surface on LinkedIn as well.

Connecting with Competitors

Everything is phenomenal about this potential Contact: you know one another, and you value their First-degree Connections. The only problem? The Connection happens to be a rival to your business or job search. Is it better to include competition into your LinkedIn network, or to avoid it at all costs?

It's a complicated consideration. On the one hand, granting them the privilege of seeing your First-degree Connections could be akin to giving ammunition to the enemy. On the other hand, by accepting a rival's Connection request, you become privy to their network and Connections as well. What if you could accept a request from a competitor, while at the same time safeguarding access to your valuable networking Connections?

A potential solution is at your fingertips here. Head to "Account and Settings" at the top-right of the LinkedIn screen, and click "Privacy and Settings." Look down to "Privacy Controls." Hit "Select who can see your connections." You can choose "Only you" as a way to ensure complete privacy of your network.

If this option seems too extreme, and you don't value the opportunity for direct reconnaissance, you're probably better off rejecting the Connection request.

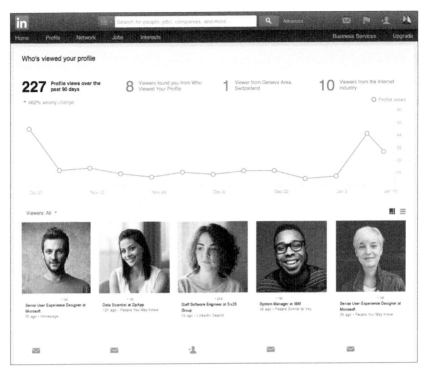

FIG. 21

The Importance of "Who's Viewed Your Profile"

Yet another valuable feature of the network, "Who's Viewed Your Profile" (see fig. 21) sheds light on the individuals who have taken glances at your personal LinkedIn hub. You can see the number of Profile views over the last 90 days, how visitors have found your Profile (Google Search, the LinkedIn Mobile App, etc.), and the companies and job titles associated with your Profile visitors. LinkedIn will also suggest Profile section additions and recommended Connections to enhance the quality and quantity of Profile views you receive.

Finally, you can check how your Profile ranks against others in your network in terms of overall views. Upgrading to LinkedIn Premium, the paid service, grants more in-depth access to "Who's Viewed Your Profile."

You can find "Who's Viewed" in two ways: first, by mousing over "Profile" at the top of the homepage, and then clicking on the feature. Second, you can head to your headshot at the top of the homepage, then hit the number near "X [number of] people viewed your profile…" When it comes to First-degree Connection Profile visitors, you can send a message or see their Profile. Fully-named second- or third-degree Connections can be invited to connect, or you can check out their Profile. Some LinkedIn members choose to be anonymous and, therefore, cannot be sought after through "Who's Viewed Your Profile."

Job seekers can use "Who's Viewed" to get a leg up on employment competition. Ensure your Profile sits on top of the pile of Profiles recruiters encounter by directly looking at the Profiles of recruiters who have checked you out. Doing so will improve the likelihood that they'll return to see your Profile.

Businesses would be wise to consider "Who's Viewed" as a warm-lead generator. Those who visit your Personal Profile or Company Page are potential clients interested in your service or product. Feel free, if possible, to extend to your profile viewers an invitation to connect or an InMail, provided it doesn't sound like a sales pitch.

Everything About the "People You May Know"

After "Who's Viewed Your Profile," "People You May Know" (see fig. 22) is cited as one of the more valued features on LinkedIn. A proprietary invention of LinkedIn, PYMK essentially predicts Connections that would make natural and worthwhile expansions of your network. At first glance, the technology can come off as a bit frightening; PYMK

People you may know

Jeff Hester
Publishing Professional
Connect • Skip

Justin Forbes
Fine Artist/Illustrator
Connect • Skip

FIG. 22

can retrieve Connections that you thought were lost and forgotten long ago. But LinkedIn's users have truly embraced the feature, as it's been responsible for nearly half of all Connections formed throughout the network. The PYMK feature is inescapable on the LinkedIn platform: look for it throughout your time on the site, and evaluate the Connections it proposes to you.

Referrals on LinkedIn

What's the good of interacting on LinkedIn if your time can't generate worthwhile off-network opportunities—whether they are interviews, sales leads, or referrals to your product, service, or expertise? Here are some strategies that can be useful in the hunt for lead- and referral-generation on LinkedIn:

Create rich media directly aimed at, and based on the needs of, your target (recruiter, company, customer, or client). This is an important point to reiterate. Creating a 30-second introductory video, a slideshow, a presentation, a webinar, an infographic, a white paper—any high-value content that proves why you are a great potential Connection—will immediately boost the referrals you attract.

Survey your Second-degree Connections on a daily basis. This means heading to your First-degree Connections' profiles and mining them for potential Contacts that might be of value to you. Maintain a list of these Second-degrees (and the First-degrees through whom you'd Contact them); it's best to prioritize those Connections who have made a Recommendation or Endorsement on the First-degree's Profile. Then, make a point to connect to that First-degree off the network, through a phone call or e-mail, asking for a Recommendation and potential invitation request to your desired Second-degree Contact.

Write one new Recommendation for a First-degree Connection every day. It takes just minutes to compose a thoughtful and thankful vote of confidence for someone in your network. Doing this establishes two goals: it testifies to your relationship and its value, and it advertises your services to your Contact's network. Someone from his or her network could passively notice the new Recommendation and Contact you for more information. Alternatively, your Contact may consider giving you a Recommendation or referral in a "return the favor" gesture down the line.

The Art of LinkedIn Recommendations

Recommendations can do more than just create referral opportunities for you. They also serve as an indicator of the health of your professional network. Remember, it's both quality and quantity that count in your LinkedIn Recommendations. Having too few of them can lessen the perception of your expertise in your industry; but too many of them can give off a quid pro quo vibe and seem inauthentic. Anemic, generic, canned Recommendations do no justice to your professional standing. Your goal should be to attain as many impressive and detailed Recommendations as you can. Here are some tips on asking for Recommendations:

Look to the workplace and beyond. Work-related Recommendations are vital, of course, but also consider volunteering experiences you've had, organizations and associations of which you are a member, and your coworkers and colleagues as sources for potentially golden Recommendations.

Clarify how quick it can be to write a Recommendation. In your message to a prospective Recommendation writer, highlight that it should take no longer than ten minutes, and three paragraphs, to accomplish the task.

Give them a leading line or two. Jog their memory with topics and circumstances that they could mention in the piece.

Suggest why you stand out. What are your professional attributes you would like the Recommendation to highlight? Politely make note of these in your request message.

Getting Introduced on LinkedIn

Much of the power of LinkedIn lies in the Second-degree Connections that your First-degree Contacts open up to you. But accessing these Connections requires more work and etiquette than randomly sending default intro-request messages to your Connections. A well-crafted message distinguishes the most successful introduction requests on this professional network (see fig. 23).

Begin with a subject line that immediately and politely piques attention. Be personal, and even a tad humorous, with this opener. Next, mention the specific way that you and your Contact know one another, ideally away from LinkedIn. Go into concise yet persuasive detail as to why you are interested in being connected to a member of their network. How can they help you achieve what you need? And add dollops of polite language throughout the message: "It would be so kind of you to . . . ; Thank you very much for . . . ; I really appreciate your helping me with . . . "

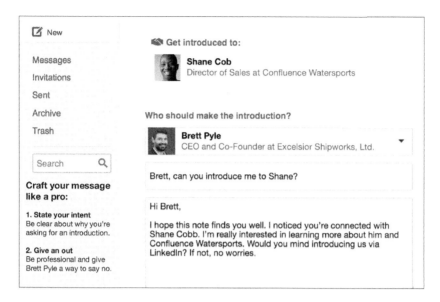

FIG. 23

To take this all a step further, include an "out" line, which basically says, "If you're uncomfortable with this introduction, no problem. Would you have any advice on another way in which I could connect with [him or her]?"

Once you have sent your introduction request, leave it alone. If you haven't heard anything after one week, feel free to extend a short and gentle reminder message. But only do this once.

Regardless of whether the introduction takes place or comes to anything of value, provide your Contact with a note of hearty thanks for their willingness to support you.

Just like in the offline professional world, a phone call seeking "a friend of a friend" is not guaranteed to create a sale or job opportunity. But in LinkedIn's streamlined world of "friends of friends," there's no better way to expand your high-value Contact base online than through these introductory requests.

How to Grow Your LinkedIn Connections

Aside from referral-generating activities and sending introduction messages, there are many other ways for you to maximize the quality of your LinkedIn network. Take advantage of these timely opportunities to expand your pool of LinkedIn Connections with actions both on and away from the network:

Solicit a LinkedIn Connection from your upcoming appointments. Whenever you schedule a meeting, lunch, coffee, or networking get-together—any type of appointment, really—immediately send an invitation to connect.

Add Contacts as soon as any networking event concludes. Your hobnobbing offline, at conferences, panels, trade shows, and similar events, is a perfect opportunity for building your LinkedIn Contact base. Just be sure to include a thoughtful, personalized message to each new Contact.

Return to "Who's Viewed Your Profile" for Connection opportunities. If someone of potentially great value scouts your Profile, reach out with an invitation.

When you connect with someone on Facebook, Twitter, Instagram, etc. Immediately send a LinkedIn request as soon as someone adds you to another social network. The window of openness to such an invitation, which requires that you be top-of-mind, can close very quickly, so capitalize on this moment.

Look to your daily online reading for potential Connections. Read an amazing or super-helpful article, e-book, white paper, or the like? Even better, did you come across this content through LinkedIn (via your Profile, LinkedIn Pulse, or LinkedIn Publish)? Like and positively comment on the post.

How Many LinkedIn Connections Should I Have?

This question has been the center of perennial debate among networking experts. There are pitfalls and advantages to both small and large networks. And the answer to the question is really one that varies depending on both the situation and goals of each individual user.

The only LinkedIn users who tend to truly benefit from hundreds of Connections, varying wildly on the familiarity scale, are recruiters who need access to as wide a pool as possible to fill employment gaps, so that they can secure a regular stream of commissions for their work, and freelancers who might get work from a varied and constantly evolving list of clients.

Too large a network can devastate job seekers, however. Having hundreds of "cold" Connections—Contacts who don't know you, can't vouch for you, and would therefore be very unwilling to let you tap into their own networks—can easily destroy a search for great employment. It's true that more Connections, even cold ones, can lead to discovering a higher number of available jobs. Still, a largely "cold" network backfires when it comes to securing interviews and job offers. On the other hand, a "warm" network, laden with Connections who know, trust, and like you, pays off in terms of establishing the credibility that recruiters and employers crave in job applicants.

The same rationale applies to LinkedIn users wanting to network or generate sales leads from the site. A "warm" network easily generates the credibility—Recommendations, Endorsements, successful introductions, etc.—needed to secure clients, sales, and high-level networking Contacts. If you can secure thousands of "warm" Connections like this, you're in great shape. But having even 50 warm Contacts is better than a network of 1,000 cold ones.

When to Get Rid of a LinkedIn Connection?

You should regularly prune your First-degree Connections for the removal of any Contacts that are not of optimum value for your LinkedIn experience, ultimately aimed at getting a job, building your networking cachet, or generating leads for your business. Prime candidates for removal include Connections:

- Who have never responded to your messages
- With networks that sprawl into the hundreds or thousands and yet have absolutely no value
- Who are unwilling to let you into their network Contacts
- Who spam you
- With whom you want to terminate any affiliation

So you've built up a solid list of useful, First-degree Connections. Now what? In the next chapter, we'll look at strategies you can use to build and maintain relationships with those new (and potentially useful) Connections.

GETTING THE MOST OUT OF YOUR CONNECTIONS

So, you've amassed a pile of valuable Contacts. What now? How do you get the most out of your First-degree Connections and friends of friends? In this chapter, we'll look at a few different strategies you can leverage to get the most out of your Connections.

How Sorting and Filtering Your Connections Helps You Win on LinkedIn

Sorting and filtering are most useful for quickly locating each and every Contact related to a particular company, marketing campaign, regional job search, or other specific categorization query you have in mind. To handle such a scenario, you would need only to sort your Contacts by

First Name, Last Name, Recent Conversation, or New—a simple and fast procedure. When you filter Contacts that you have already sorted, you further fine-tune your search: by First-degree Connections, by company, by location, by saved search, and a range of other options.

You can also use filtering to stunning effect to mine the Connections of your First-degree Contacts, searching for candidates for future introduction requests. First, type the Contact's name into the search bar, but DO NOT hit "Enter"; instead, click the magnifying glass search icon. This takes you to a search results page, where you then click on the number indicating your Contact's connections. This will open a page revealing all of your chosen Contact's Connections. On the left side of the screen, you will be able to filter this result for Second-degree Connections. LinkedIn Premium members will be able to search by other potentially attractive categories, such as Years of Experience, Seniority Level, Fortune 1000, and Recently Joined Members.

Effective Messaging on LinkedIn

Here are a few tips to employ when crafting messages for invitation requests, InMails, and beyond:

Always create a customized, catchy subject line and body. Personalize what you write to specifically intrigue and connect with the unique Contact to whom you're writing. Do your best to ensure the subject line will persuade the recipient to open the message.

Get creative with the 300-character limit. For example, even though LinkedIn doesn't allow you to formally include a link in the message, you can skirt around this with some clever rewriting (www . visitme . org— note the added spaces).

Include a "call to action." Also known as a "CTA," this component is a verb-based method of keeping the Connection going between you and your message recipient. "Visit my website," "Let me know when might be best to meet," "Check out this blog post—it might help you out"; these are all great CTA examples.

Always answer the question, "Why are you getting in touch with me?" Both your subject line and the message itself should place your desire to connect in clear and specific context.

Keep it quick. Take no more than the first one to two sentences to mention what your goal would be in connecting with this individual.

Be grateful and be gone. Conclude the message with a line that communicates how thankful you are for the opportunity to connect. Signing off with "thank you" is always a classy touch.

Effective Tagging on LinkedIn

A "tag" is a piece of text used for the purpose of identifying and classifying a Contact on LinkedIn (see fig. 24). Tags can be as long as 100 characters, and can be applied to any person in your Saved Contacts on LinkedIn, regardless of whether or not they're in your network. Tags are also private; only you can see how you've tagged Contacts and Contacts-to-be on the site. (Tag a Contact by clicking "Relationship" under the photo of the member you wish to tag, then clicking "tag" and adding a tag. Remove a tag by unchecking the box next to it.)

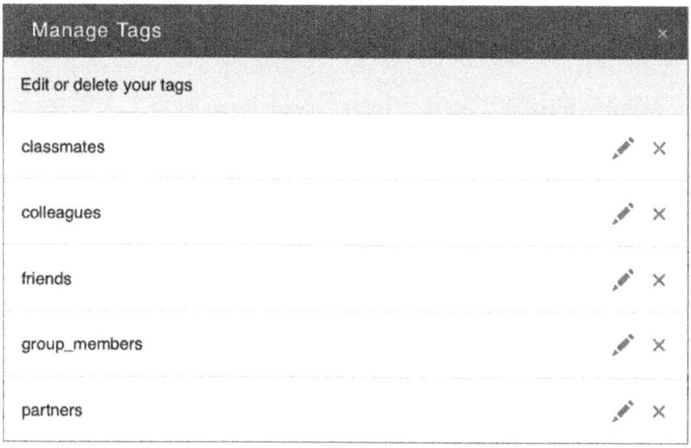

Manage Tags		✕
Edit or delete your tags		
classmates	✎	✕
colleagues	✎	✕
friends	✎	✕
group_members	✎	✕
partners	✎	✕

FIG. 24 Source: http://www.linkedin.com

Tagging Contacts on LinkedIn can dramatically simplify your sorting, filtering, and messaging activity. You can use tags to remind yourself of how you and a person connected, made easy by the tag's appearing on the right-hand side of the screen when viewing a Contact. You can also send a mass message (not always recommended, however) to all those Contacts with a particular tag that you create: "New York City," "mobile marketing," "distributors," etc.

The Art of Relationship–Building on LinkedIn

Using all of the tools at your disposal on LinkedIn can be a bit overwhelming. Orchestrate your efforts in the pursuit of better relationships on the network, through all the valuable features and assets available to you, using the following suggestions:

Respond promptly. Whenever you receive a message or Contact of any kind on LinkedIn, do your best to reply to it right away. (This is greatly simplified when you're on the go through the use of LinkedIn's mobile phone app.) You don't necessarily need to have the perfect solution or answer to their message immediately. But even saying "Thanks for your message, let me get back to you on this as quickly as I can" communicates that you care, that you are actively seeking a solution, and that you prioritize your relationship with the Contact. All of these qualities reinforce winning high-value LinkedIn Connections for the long term.

Solve the problem, answer the question, or provide a solution in one day, maximum. Clarifying that you received the message and will reply ASAP is the first step, but it requires follow-through. Ensure that you provide a helpful response to the message within 24 hours at the most. This behavior sends a message of reliability and conscientiousness to your Contact, which is sure to earn and galvanize his or her trust and support. Job seekers will see this tip result in a higher response rate and increased quality communication with potential employers, recruiters, recommendation writers, and more.

Respect the maxim that the Contact, or the client, is always right. Provide every customer, job lead, and networking Contact with the royal treatment. Be polite, helpful, non-sales, and value-first in all of your communications on LinkedIn. Who you might perceive to be small-fry or insignificant clients and connections today may develop into the industry-leading pioneers and seven-figure-value customers of the future. At the very least, every Contact can potentially introduce or refer you to someone else. Being part of the professional development of your network members, through thoughtful and speedy communication with them, attracts great LinkedIn karma and can do nothing to harm your own ascent in your industry.

Speak to or meet with your Connections and clients often. Make an effort to communicate with 5 to 10 members of your network, ideally offline, at least twice a week. This can take place as a phone call or Skype connection, a meeting over coffee, or at a networking event. Be sure to avoid discussing "business" per se, instead concentrating on topics that will lend value to your Contact: articles you've read that might help them, people you'd like to introduce them to in your network, the latest industry trends and news that would appeal to them. This applies to job seekers and businesses alike. Communication in this way maintains the "warmth" of your network and makes it much simpler to secure an intro-duction, recommendation, referral, or repeat purchase.

Boost the quality of your LinkedIn Groups participation. LinkedIn's Groups regularly earn acclaim on the web for their usefulness in forging valuable connections between industry professionals. Your participa-tion in the Groups most relevant to your sector and interest can lend you thought-leading credibility and authority. Regularly participating in Groups establishes trustworthiness with other members, who may be actively or passively viewing your Profile or Company Page and considering making Contact with you. Aim to become a Top Contributor in each group by commenting and creating valuable discussion posts on a daily basis.

Use Keyword-rich saved searches. Saved searches are the most reliable way to establish a steady stream of valuable business and relationship-building opportunities (see fig. 25). Ensure that every saved search you create is based on a target Keyword or phrase directly relevant to your professional needs and aspirations. Determine who you most need to form relationships with both today and in the future. Seeking a job in animation on the West Coast? Create a search for "animation art California," for example. A business in the construction industry might do well to regularly search for "strategic planner construction design."

Capitalize on daily status updates. Crafting a useful status update at least once a day communicates volumes of value to your network's members. Have you found a useful new app or website? A must-read book, article, or blog post? Just bought a pass or ticket to an awesome networking event? Did a nifty, practical tip come to your attention recently? Have you rubbed elbows with anyone of note in the past few days? What question would you love to field answers for from your network? All of these are great ideas for status update prompts.

Make use of LinkedIn Contacts. Found on both the official LinkedIn website, as well as the network's mobile app, LinkedIn Contacts essentially supercharges your relationship-building activities throughout the site. It centralizes all of your Contacts' information in one handy location, synchronizing this with details on the conversations, notes, and meetings for each of your Contacts. Contacts notifies you all the time: when it's been too long since you've last reached out to someone, when a Connection is starting a new job, etc. And you can customize your own reminders through the feature as well.

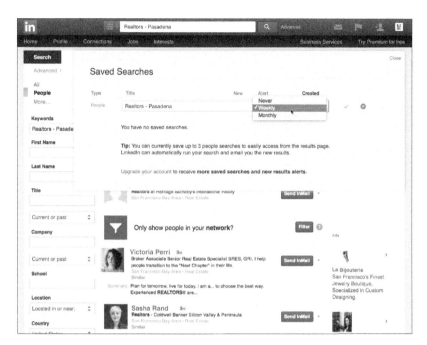

FIG. 25

Regularly recommend and endorse your Connections. At least once
a week, spotlight someone in your network who merits recognition
for being as great as they are. Be thoughtful in how you recommend
or endorse. Select only a handful of the most on-brand skills that are
the best match for this Connection. Include specific details in your
Recommendation that prove how well you know the Connection and
why he or she is so valuable. These actions can only be returned to you
in positive relationship energy: not necessarily in a swap-for-swap
Recommendation or Endorsement, but potentially in the form of an easier invitation request or improved offline communication down the line.

Final Thoughts: You Get Out What You Put In

Over the course of this book, we've taken a good look at the ins and outs of LinkedIn. From the Personal Profile to the Company Page, to Groups, searching, and connecting, no proverbial stone has been left unturned. Still, even after having read this whole text, and even with LinkedIn as a great and powerful tool, your professional networking isn't going to create itself.

Like any sort of networking or journey for professional advancement, the benefits you reap from LinkedIn will strongly correlate with the amount of time and effort you put into it. If you take the time to build up a complete Profile, gain useful First-degree Connections, and consistently engage in a meaningful and positive way, then LinkedIn can truly be a wellspring of information and opportunity. Without that sort of commitment—if you just set up a Profile and forget about it—what could otherwise be a hugely useful tool can turn into just another website sending you daily e-mails to delete. You now have the tools—how you use them is up to you.

INDEX

CPSIA information can be obtained at www.ICGtesting.com
Printed in the USA
BVOW11s0558301015

424767BV00012B/13/P